Hands-On Projects

JUNGLE SAFARI

Active Learning about the Environment

by Carol Wawrychuk & Cherie McSweeney
illustrated by Philip Chalk

Contents

For a complete catalog, please write to the address below:
P.O. Box 1680, Palo Alto, CA 94302 U.S.A.
Call us at: 1-800-255-6049
E-mail us at: MMBooks@aol.com
Visit our Web site: http://www.mondaymorningbooks.com

Monday Morning Books is a registered trademark
of Monday Morning Books, Inc.

ISBN 1-57612-072-4
Printed in the United States of America
987654321

Introduction

Tigers, lions, giraffes, zebras, and hippos! These are just some of the wild animals children will meet during their exciting journey into the jungle! Although these creatures live in different parts of the world, you'll find all these animals here, in the *Jungle Safari* unit.

Introduce children to the jungle environment with a photo safari. Children slip into their safari vests and hats (made from a pillowcase and butcher paper). They prepare to take pictures with cereal box cameras. The journey continues over rugged landscapes in an all-terrain vehicle, or winding down river rapids in a rowboat. Both of these vehicles are created from large boxes. Along the way, the children can photograph an appliance-box hippo and wave to a sawhorse giraffe!

Children learn about the jungle as they design a mural filled with jungle themes. They imagine their jungle journeys when they draw pictures (representing the "photos" they've taken). Through counting and matching activities, young learners become engaged in cognitive thinking.

The *Jungle Safari* unit opens a colorful world of plants and animals to children. So grab your camera, hop into the all-terrain vehicle, and let the adventure begin!

Helpful Hints from the Authors:

- Stores often will give away their empty boxes.
- Boxes can collapse for easy transporting and storing.
- Some boxes may be reused for different projects.
- Use cable ties, string, or yarn for assembling the box projects. (Cable ties provide the sturdiest form of attaching.)
- A mat knife or Exacto blade works best for cutting boxes, but a sharp kitchen knife will work well, too. Use the knife or blade well out of the way of children!
- Local businesses will often donate supplies.
- Shop discount stores for tongue depressors, paper cups, sponges, and paper goods.
- Involve parents in projects. They can save items for you.
- Provide globes and atlases for children to use to locate real jungles around the world!
- Consider taking children on a field trip to a local zoo or wildlife park as a finale to this unit!

All-Terrain Vehicle

Materials:
Large box (with flaps on the ends, not the sides), large black plastic deli trays, construction paper (white and red), cable ties, brown tempera paint, shallow tins for paint, paint rollers, glue, scissors, sharp instrument for cutting (for adult use only)

Directions:
1. Tape the box closed and place it on its side. Remove one half of the top side of the box according to the diagram below.

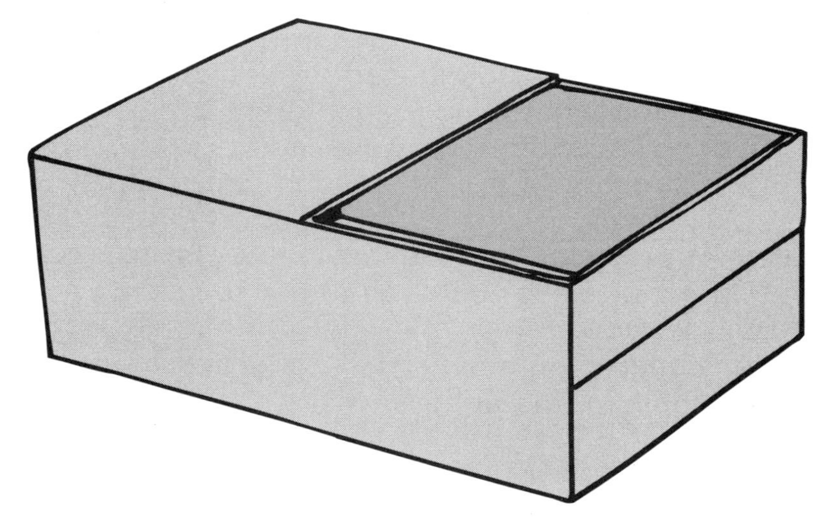

All-Terrain Vehicle

2. Cut a windshield and doors according to the diagram.

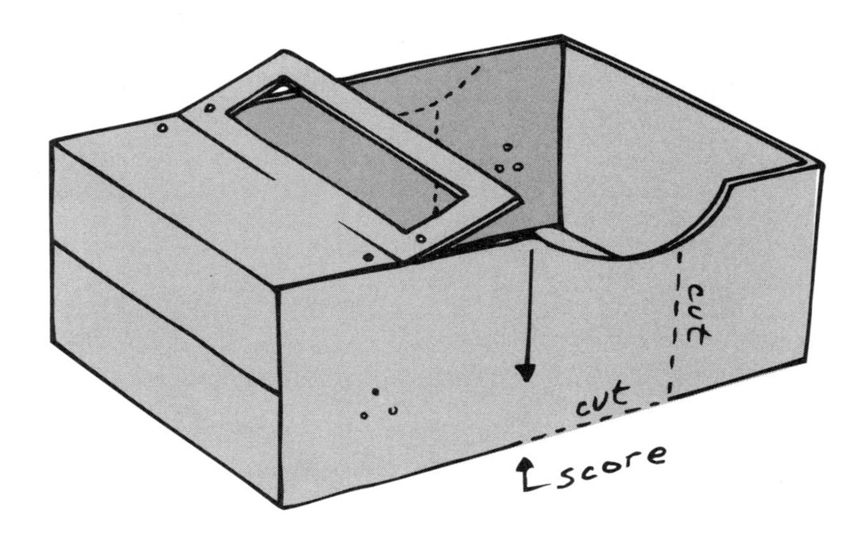

3. Poke holes in the box to secure the windshield and tires.
4. Have the children use rollers to paint the box.
5. Once the paint has dried, flip up the windshield and secure it to the upper portion of the box with cable ties.
6. Poke holes in the deli trays and align these with holes on the sides of the box. Attach the wheels with cable ties.
7. Have the children cut out headlights and taillights from the construction paper and glue these to the vehicle.

Options:

- If deli trays are not available, cut wheels from construction paper, or paint wheels directly onto the vehicle.
- Cover the windshield with clear cellophane.

Video Link:

- *Really Wild Animals—Swinging Safari* (National Geographic Kids Video)

Rowboat

Materials:

Paddle Pattern (p. 7), large appliance box, two wrapping paper tubes, sturdy paper, masking tape, tempera paint, shallow tins for paint, large sponges, marker, scissors, sharp instrument for cutting (for adult use only)

Directions:

1. Tape the box closed and place it on its side.
2. Cut off one long side of the box. Cut a curve into opposite sides of the box. Secure the inner flaps with masking tape.
3. Cut two slits at one end of each wrapping paper tube.
4. Trace the paddle pattern onto sturdy paper twice and cut it out.
5. To make the oars, insert the paddles into the slits in the tubes and reinforce with masking tape.
6. Have the children sponge-paint the boat and oars.

Video Link:

• *Really Wild Animals—Wonders Down Under* (National Geographic Kids Video)

Paddle Pattern

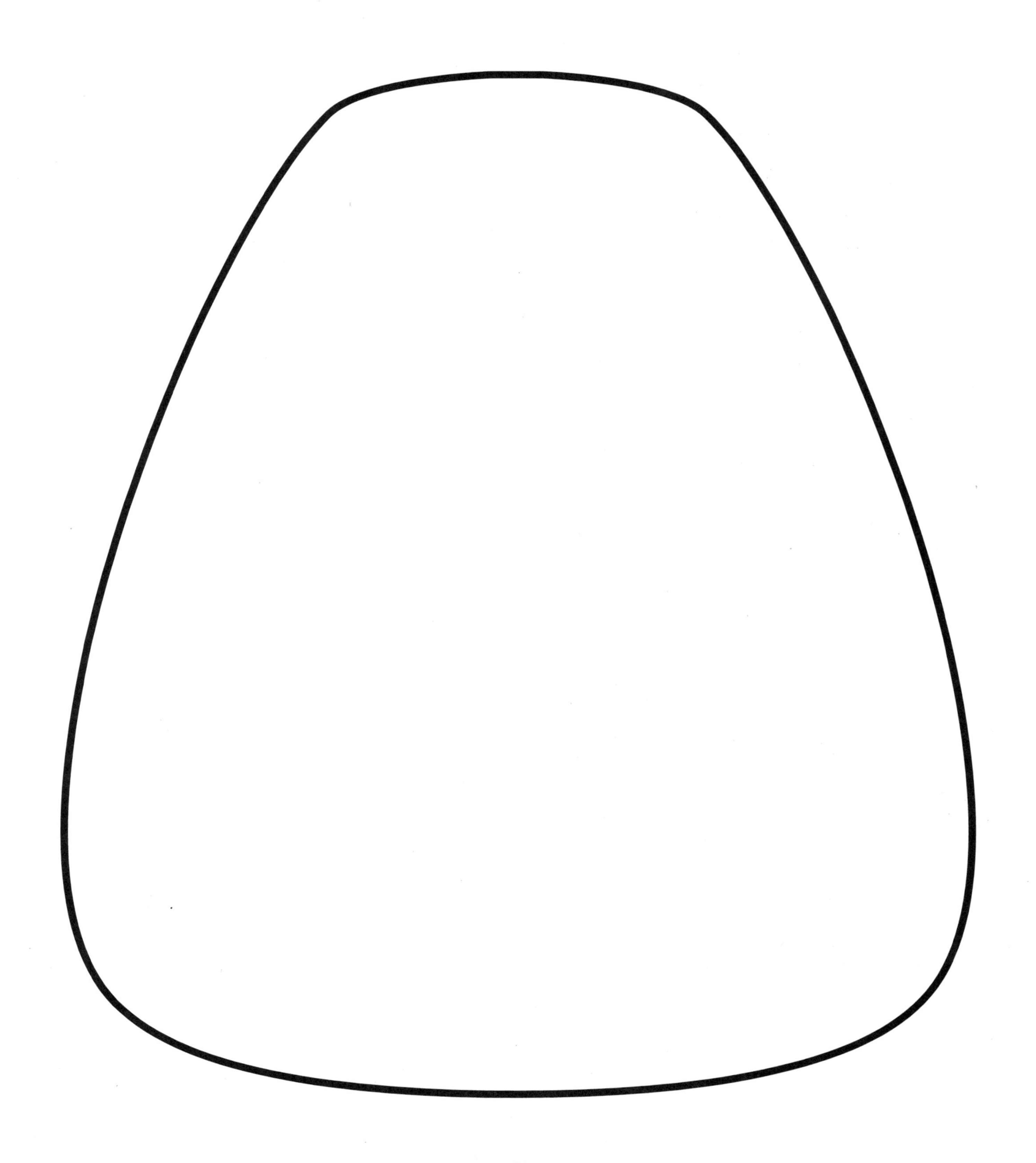

Huge Hippopotamus

Materials:

Hippopotamus Patterns (p. 10), large box (the size of a two-drawer file cabinet), computer paper box (without lid), four 39-oz. (1.10 kg.) coffee cans or large cylindrical oatmeal containers, four empty thread spools, cable ties, marker, construction paper (brown and white), masking tape, paste, glue, tempera paint (white, brown, and black), shallow tins for paint, large paintbrushes or rollers, scissors, sharp instrument for cutting (for adult use only)

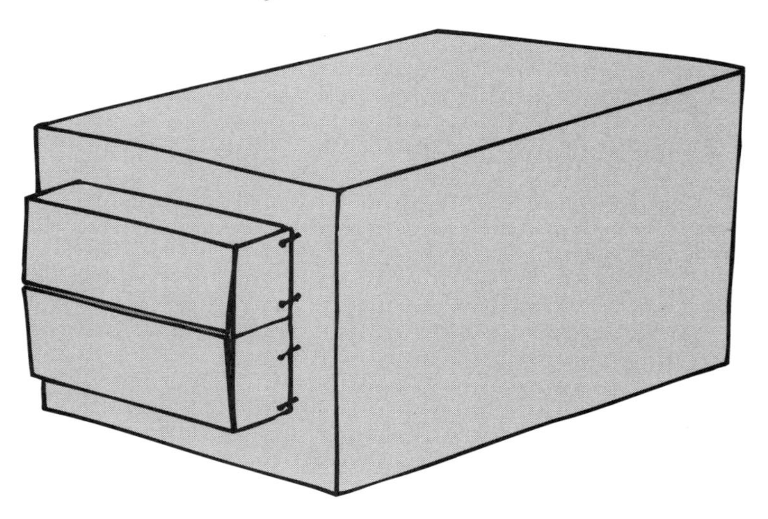

Directions:

1. Connect the open end of the computer paper box to the large box by poking aligning holes in several places of both boxes and securing with cable ties.
2. Open the flaps on the bottom of the computer paper box.
3. Trace the Hippopotamus Patterns onto construction paper and cut them out.
4. Cover the cylindrical containers with white paper.

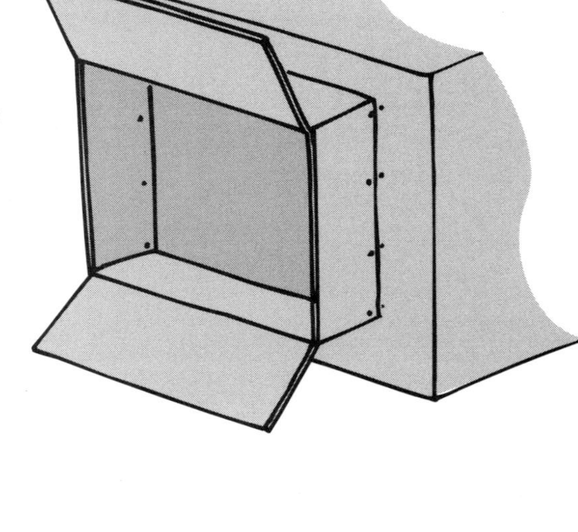

Huge Hippopotamus

5. For the legs of the hippopotamus, glue the four cylindrical containers to the bottom of the large box. Reinforce with masking tape.
6. Have the children mix white and black paint together to make a dark gray.
7. Provide paintbrushes and rollers for the children to use to paint the hippopotamus gray and brown.
8. Once the paint has dried, have the children paste the curved ears to the top of the head, the tail to the back, and the thread spool teeth to the inside of the mouth.
9. Eyes can be painted on, or cut from construction paper and glued on.

Option:

• Have the children paint the inside of the hippo's mouth pink.

Book Link:

• *Hip-Hip-Hip-Hippopotamus* by Mary Rice Hopkins (Crossway)

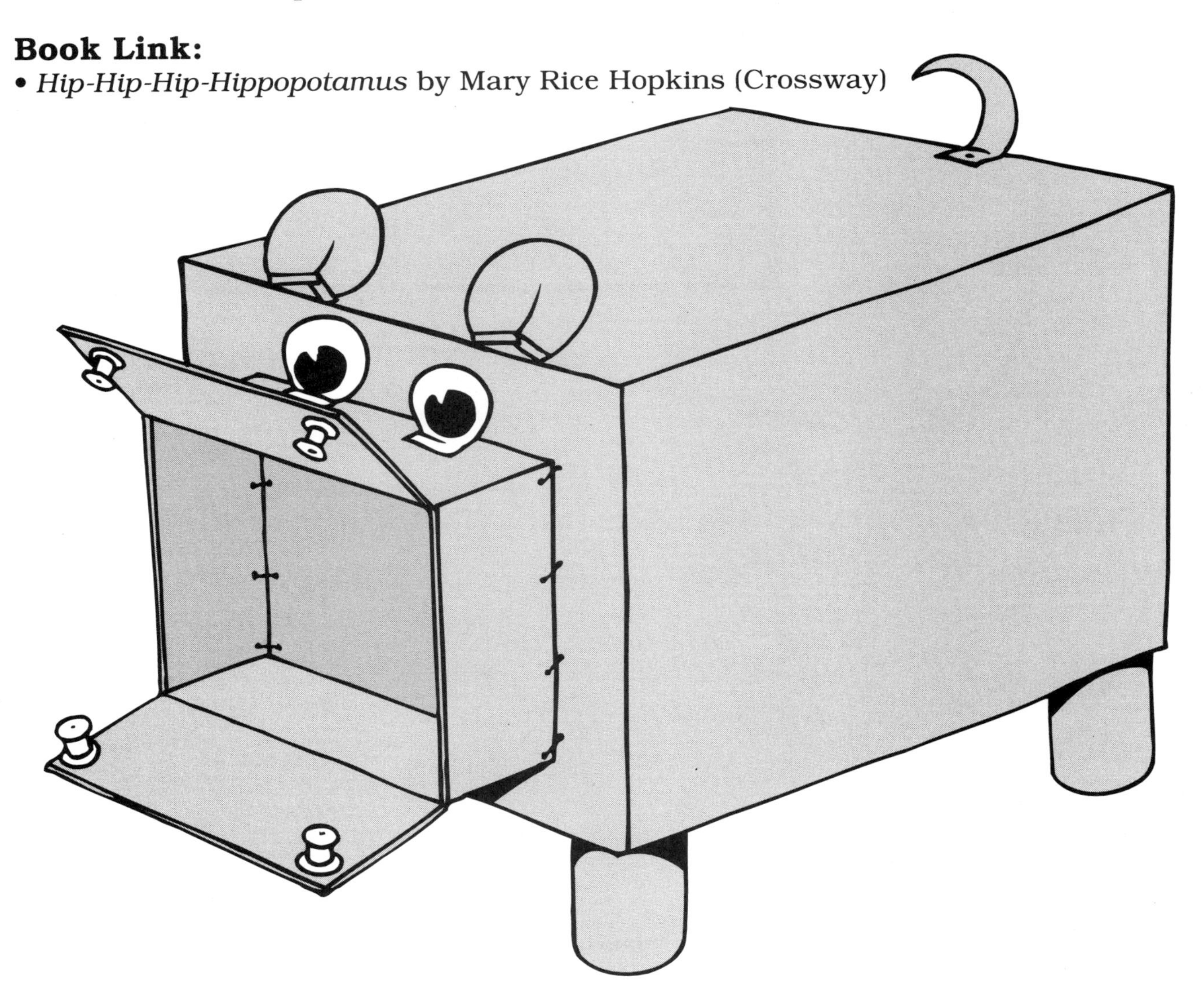

Hippopotamus Patterns

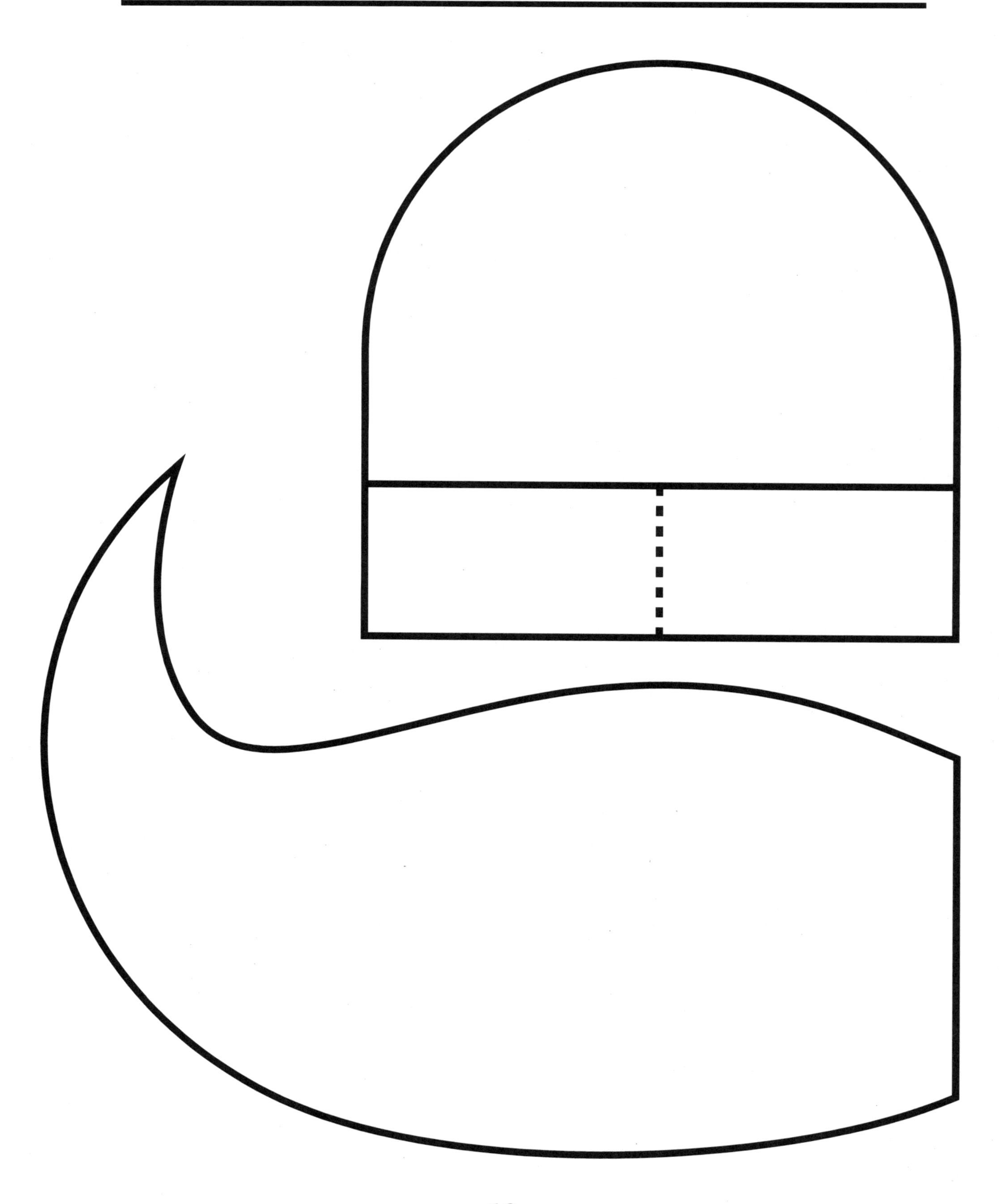

Sawhorse Giraffe

Materials:
Giraffe Head Pattern (p. 12), sawhorse, flat section of large cardboard box, pipe cleaners (brown and yellow), yarn, markers, tempera paint (yellow and brown), shallow tins for paint, paintbrushes or sponge brushes, scissors, sharp instrument for cutting (for adult use only), saw (for adult use only), hammer and nails (for adult use only)

Directions:
1. Cut the sawhorse to a height that allows children to sit on it safely.
2. Enlarge and duplicate a copy of the giraffe head.
3. On the flat piece of cardboard, trace the giraffe head twice, extending the neck (as shown), and cut out. Cut notches in the necks to allow for the sawhorse legs (as shown). Nail the heads on both sides of the sawhorse, tying them together around the ear with yarn.
4. On a piece of cardboard, draw the giraffe's tail and cut it out. Nail the tail to the giraffe.
5. Have the children paint the giraffe yellow.
6. Once the paint has dried, children can sponge-paint brown spots onto the giraffe.
7. Have the children color the giraffe's eyes and mouth with markers.
8. Punch holes in the giraffe's neck and have the children twist-tie short sections of pipe cleaners through the holes for the mane.

Options:
- Let children make handprints on the giraffe for spots.
- Glue small strips of black construction paper on for eyelashes.
- Glue strips of construction paper or pieces of yarn to the end of the tail.

Giraffe Head Pattern

Flowering Jungle Tree

Use this activity with Hanging Monkey (p. 16) and Slithering Snake (p. 19).

Materials:
Jungle Leaf Patterns (p. 15), large flat piece of cardboard, large paper grocery bags, brown tempera paint, shallow tins for paint, squeegees or soap scrunchies (the nylon-net type that come with liquid body soap), sturdy paper, green construction paper, colored tissue paper, pipe cleaners, masking tape or heavy packing tape, scissors, staplers, markers, sharp instrument for cutting (for adult use only)

Directions:
1. Cut a simple tree trunk shape from the flat piece of cardboard.
2. Cut open the grocery bags so that they lay flat.
3. Trace the Jungle Leaf Pattern onto sturdy paper and cut it out. Make several leaves for the children to use as templates.
4. Provide squeegees and soap scrunchies for the children to use to paint the tree trunk.
5. Help the children crumple, twist, and tape the grocery bags into branch shapes.

Flowering Jungle Tree

6. Have the children trace the leaf stencils onto green construction paper and cut out. They should make many leaves. When they're finished, staple the leaves together, as shown below.

7. Once the paint has dried, curve the tree trunk into a semi-circle for a three-dimensional effect, and secure it to the wall with a stapler, masking tape, or heavy packing tape.
8. Attach the branches to the trunk with masking tape. Tape the ends of the branches to the wall.
9. Staple the strips of leaves to the branches.

Directions for flowers:
1. Cut tissue paper into approximately 6-inch x 6-inch (15-cm x 15-cm) squares.
2. Have the children layer several sheets of tissue paper and gather the sheets together in the middle.
3. Show the children how to twist a pipe cleaner around the tissue paper to form a stem.
4. Attach the flowers to the branches with the pipe cleaners.

Jungle Leaf Patterns

Hanging Monkey

Materials:
Monkey Patterns (pp. 17-18), construction paper (brown or black), markers, crayons, scissors, hole punch, brads

Directions:
1. Trace the Monkey Body Pattern onto construction paper, mark the dots, and cut out. (Make one body for each child.)
2. Trace the Monkey Limb Patterns onto construction paper, mark the dots, and cut out. (Make one set per child.)
3. Provide a hole punch for the children to use to punch holes through the dots.
4. Show the children how to attach the arms, legs, and tails to the monkeys with brads.
5. Provide markers and crayons for children to use to add facial features and any other desired details to their monkeys.

Book Links:
• *From Head to Toe* by Eric Carle (Harper-Collins)
• *Curious George* by H. A. Rey (Houghton Mifflin)

Monkey Body Pattern

Monkey Limb Patterns

Slithering Snake

Materials:
Tongue Pattern (bottom of this page), pantyhose (one pair for every two children), small Styrofoam pieces, twist ties (one per child), felt, buttons, tempera paint (in assorted colors), shallow tins for paint, cotton swabs, glue, scissors

Directions:
1. Remove the nylon legs from the pantyhose.
2. Trace the tongue pattern onto felt pieces and cut out. Make one felt tongue per child.
3. Give each child one nylon leg to stuff with Styrofoam pieces.
4. Show the children how to close the ends of the nylons with twist ties.
5. Provide cotton swabs for the children to use to paint their snakes.
6. Have the children glue the button eyes and felt tongue onto their snakes.

Options:
- Children can use the Eye Pattern (below) instead of buttons.
- Crumpled newspaper can be used in place of Styrofoam pieces.

Book Link:
- *Verdi* by Jannell Cannon (Harcourt Brace)

Cereal Box Camera

Materials:

Individual-size cereal box (one per child), black tissue paper, liquid starch, shallow tins for starch, paintbrushes, sponges, metal juice lids (from concentrate juice containers), seam binding, empty thread spools, scissors, glue, sharp instrument for cutting (for adult use only)

Directions:

1. Remove the inner bag lining from each cereal box.
2. Tape the top of each box closed.
3. Cut sponges into small squares. These will serve as the buttons on the top of the cameras. Make one square per child.
4. Provide tissue paper pieces and starch for the children to use to cover the boxes.
5. Once the starch has dried, cut a flap, a viewing window, and two holes in each box.

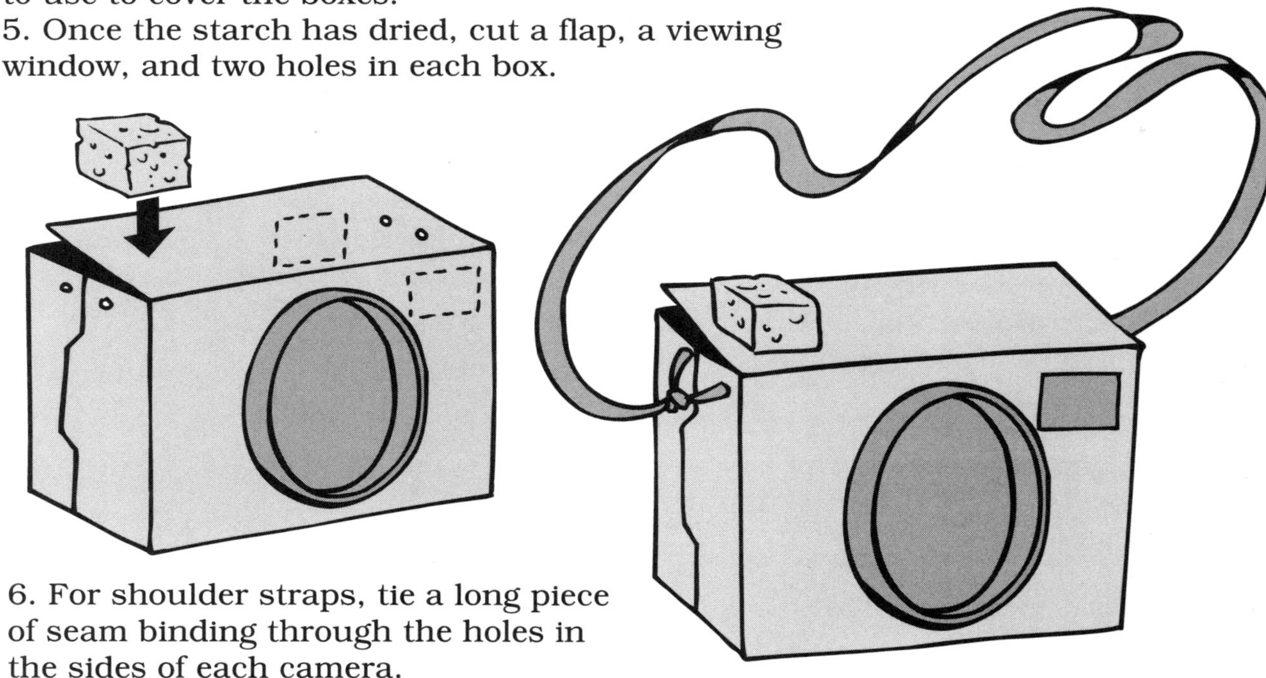

6. For shoulder straps, tie a long piece of seam binding through the holes in the sides of each camera.
7. Have each child glue a metal juice lid to the front of his or her camera for a lens and a sponge button to the camera's top.
8. Provide empty thread spools for the children to insert in the top of their cameras for film.

Option:

- Children can draw pictures representing the photos they've taken. Post these pictures in a Jungle Safari Photo Book.

Book Link:

- *My Camera at the Zoo* by Janet Perry Marshall (Little, Brown)

Safari Vest

Materials:
Standard-size pillowcases (one per child), felt (in assorted colors), ric-rac, lace, seam binding (available at fabric stores), buttons, newspaper, markers, glue, scissors

Directions:
1. Cut the pillowcase according to the diagram below.

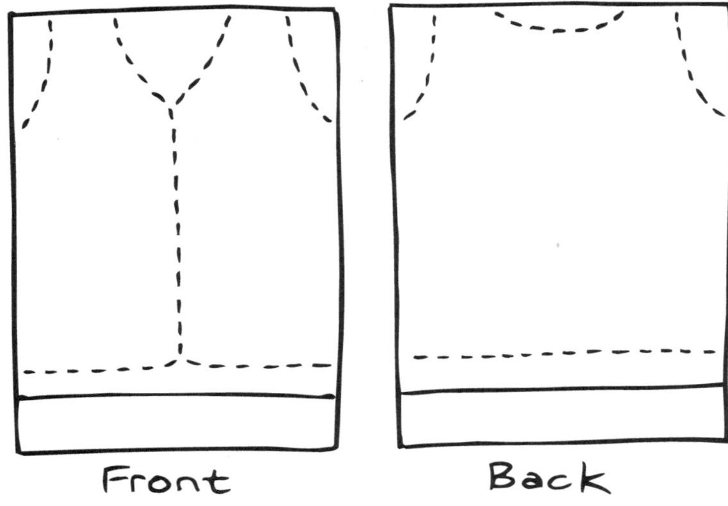

2. Cut the felt into pockets of various sizes and shapes.
3. Place each vest flat on a table and insert a section of newspaper between the front and back. This will prevent the glue from seeping through the fabric.
4. Have the children glue scraps of ric-rac, lace, seam binding, buttons, and pockets onto their safari vests.
5. Children can add desired details with markers.

Book Link:
• *That's Good! That's Bad!* by Margery Cuyler (Henry Holt)

Safari Hat

Materials:
Butcher paper, wire egg whisks, tempera paint (green and brown), shallow tins for paint, masking tape, scissors

Directions:
1. Cut butcher paper into 24-inch x 24-inch (60-cm x 60-cm) squares. Make one square per child.
2. Provide wire egg whisks for the children to use to paint the butcher paper.
3. Once the paint has dried, place each child's butcher paper on top of his or her head and mold to form a hat.
4. While the hat is on the child, make a band around it with masking tape.
5. Remove the formed hat and roll up the brim.

Options:
- Children can use animal-themed rubber stamps to decorate the butcher paper hats.
- Use colored masking tape to make hat bands.

Book Link:
- *Papa Ob Long—The Animal's Great Journey* by Leroy Blankenship (Thomas Nelson)

Safari Necklace

Materials:
Animal Patterns (p. 24), construction paper (in assorted colors), yarn, yarn needles, drinking straws, marker, hole punch, scissors

Directions:
1. Trace the animal patterns onto construction paper and cut out. Make enough for each child to have several for his or her necklace.
2. Cut straws into small sections.
3. Tie the yarn needles to long sections of yarn. Tie sections of straws to the other ends.
4. Have the children punch holes in the animals they've chosen for their necklaces.
5. Using the yarn needles, the children thread the animals and straws to make necklaces.
6. Once each necklace is complete, remove the yarn needles and tie the loose ends of yarn together.

Option:
• Instead of straws, children can alternate threading beads or macaroni with the animal patterns.

Animal Patterns

Pretty Parrot

Materials:
Parrot Patterns (pp. 26-27), construction paper (in assorted colors), colored feathers, brads, hole punch, markers, crayons, scissors, glue

Directions:
1. Trace the Parrot Patterns onto assorted colors of construction paper. (Be sure to mark where the holes will be punched.) Make a full set of patterns for each child.
2. Have each child cut out a parrot and punch a hole in both sections of the beak and the head.
3. Demonstrate how to attach the parrot's beak to the head with a brad.
4. The children can glue the eyes, wings, and tails to the birds.
5. Provide feathers, markers, and crayons for the children to use to decorate their parrots.

Option:
• Provide colored feathers for children to add to the parrots' tails.

Parrot Patterns

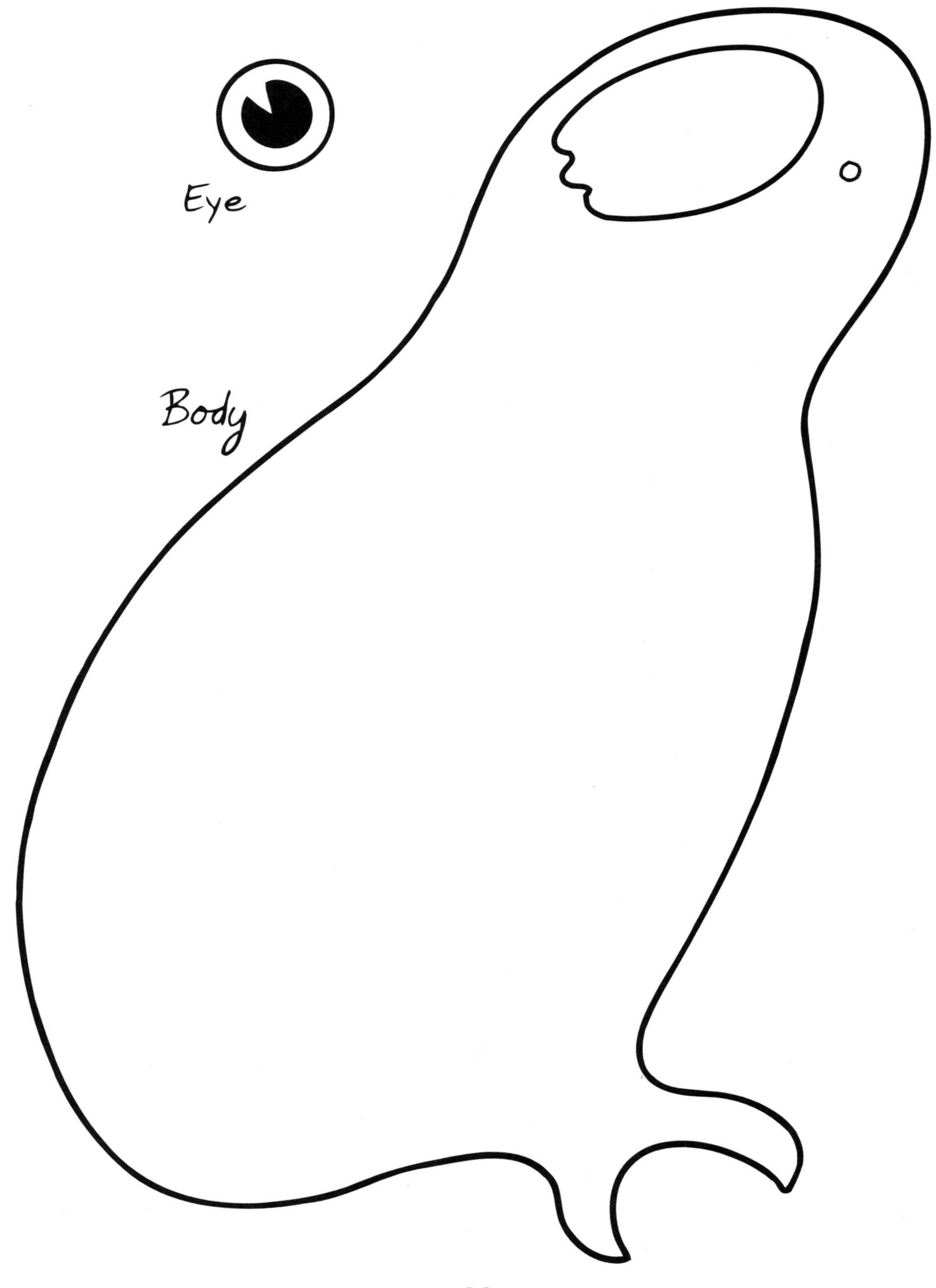

Parrot Patterns

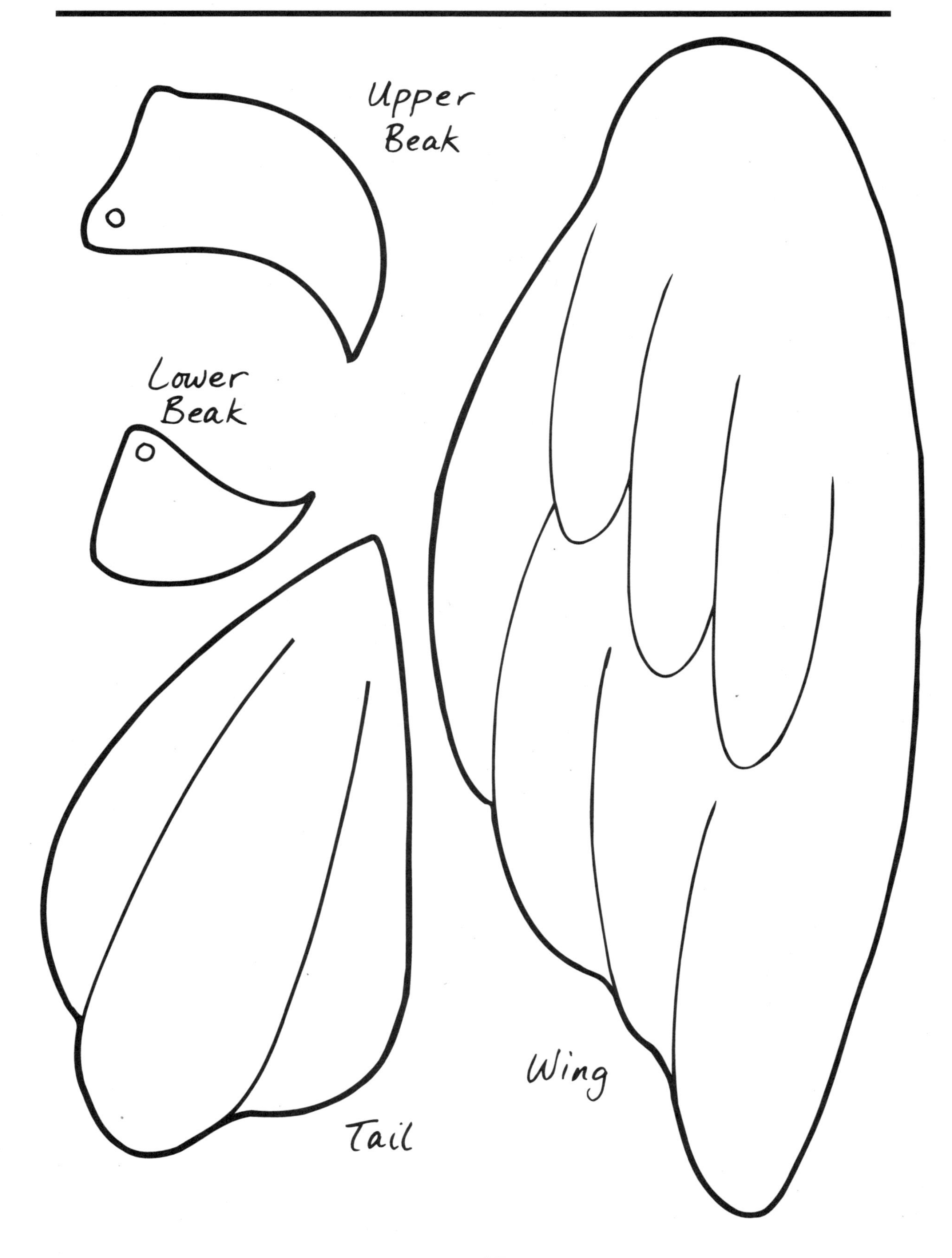

Lion Puppet

Materials:

Lion Puppet Patterns (p. 29), lunch-size paper bags (one per child), large lightweight paper plates (one per child), construction paper (yellow, brown, and black), pipe cleaners (yellow and brown), crayons or markers, scissors, glue, hole punch

Directions:

1. Punch holes around the edge of each paper plate.
2. Cut the pipe cleaners in half.
3. Trace the Lion Puppet Patterns onto construction paper. Make one set per child.
4. Have the children cut out the patterns.
5. Have the children color the paper plates using crayons or markers, then twist tie the pipe cleaners through the holes to make the lions' manes.
6. Demonstrate how to glue the puppet head to the flap of the paper bag. Let the puppets dry.
7. Have the children glue the facial features and body parts to the lions.

Book Link:

- *Dandelion* by Don Freeman (Viking)

Lion Puppet Patterns

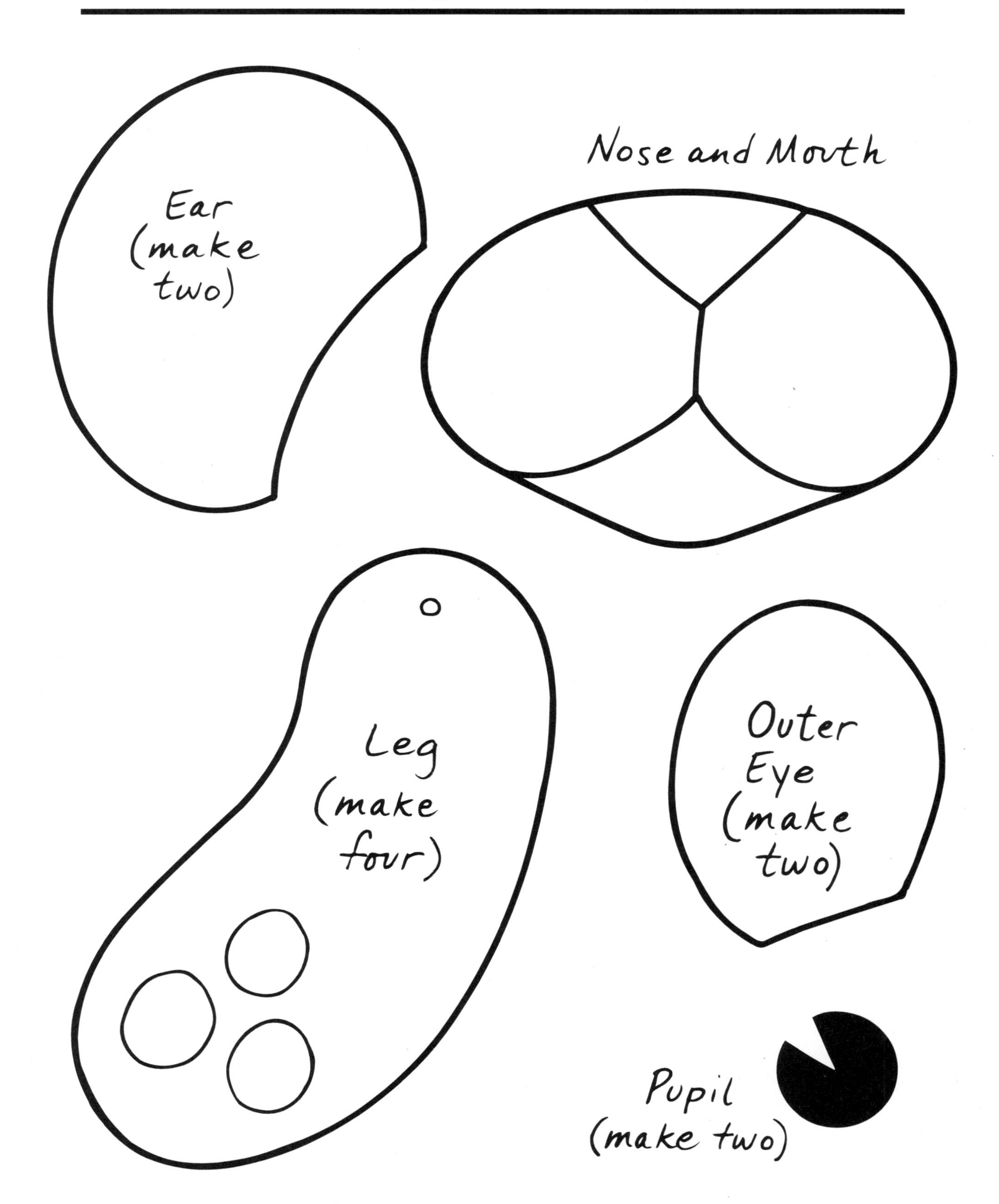

Paper Plate Elephant

Materials:
Elephant Patterns (p. 31), Styrofoam meat trays (one per child), large lightweight paper plates (one per child), sturdy paper, gray construction paper, wiggly eyes or self-sticking dots, tempera paint (black and white), shallow tins for paint, paintbrushes, glue, markers, brads, hole punch, scissors

Directions:
1. Trace the Elephant Patterns onto sturdy paper and cut them out. Make several for the children to use as templates.
2. Punch holes in each paper plate and meat tray according to the diagram below.

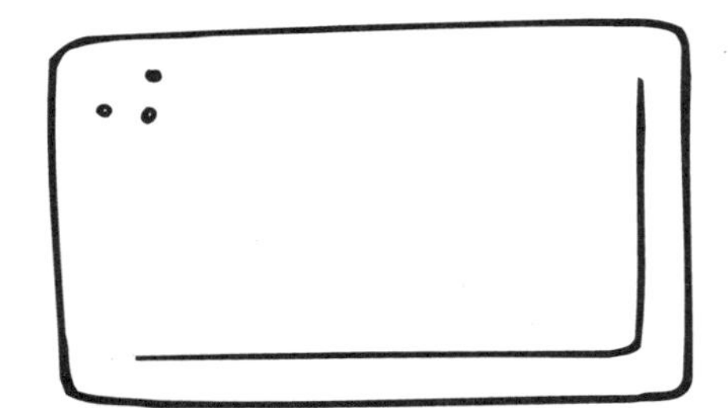

3. Have the children attach the paper plates to the meat trays with brads.
4. Let the children mix black and white paint together to make gray. Add a little glue to the paint mixture.
5. Have the children trace the patterns onto gray construction paper and cut them out.
6. Provide paintbrushes for the children to use to paint the paper plates and meat trays.
7. While the paint is still wet, have the children attach their elephant's ears, trunk, legs, tail, and wiggly eyes. (If using self-sticking dots for the eyes, attach after the paint has dried.)

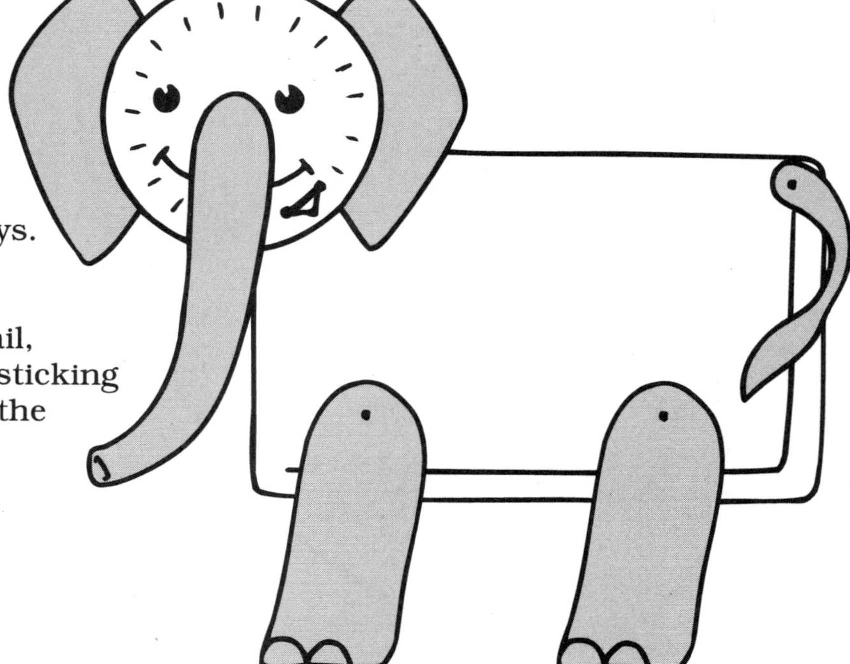

Elephant Patterns

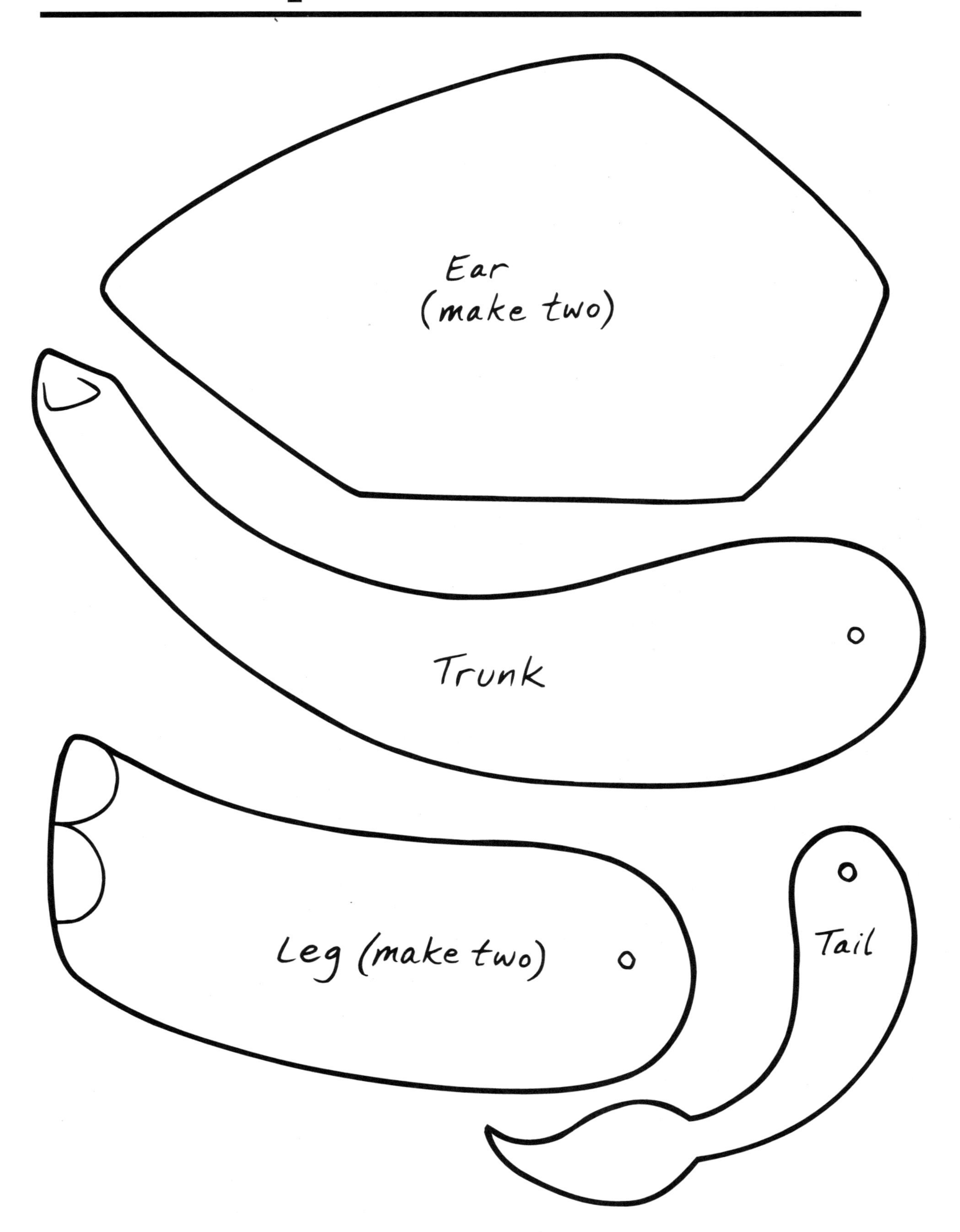

Jungle Safari Mural

Use this activity with the Handprint Butterfly (p. 33), Stand-Up Tiger (p. 34), Zany Zebra (p. 36), Coffee-Covered Giraffe (p. 38), and Colorful Fish (p. 40).

Materials:
Butcher paper, tempera paint (in assorted colors), a variety of painting utensils, shallow tins for paint, crayons and markers, scissors

Directions:
1. Cut a piece of butcher paper to the desired mural length.
2. Help the children design a jungle scene. If they need help thinking of ideas, suggest that they include hills, rivers, lakes, sky, clouds, sun, and trees.
3. Provide tempera paint, painting utensils, markers, and crayons for the children to use to create their mural.

Book Link:
• *Animals of the Jungle* by Teresa O'Brien (Flying Frog Publishing)

Handprint Butterfly

Materials:
Toilet tissue tubes (one per child), construction paper, tempera paint (in assorted colors, including black), small paintbrushes, pipe cleaners, marker, scissors, sharp instrument for cutting (for adult use only), tape, clear fishing line or yarn (optional)

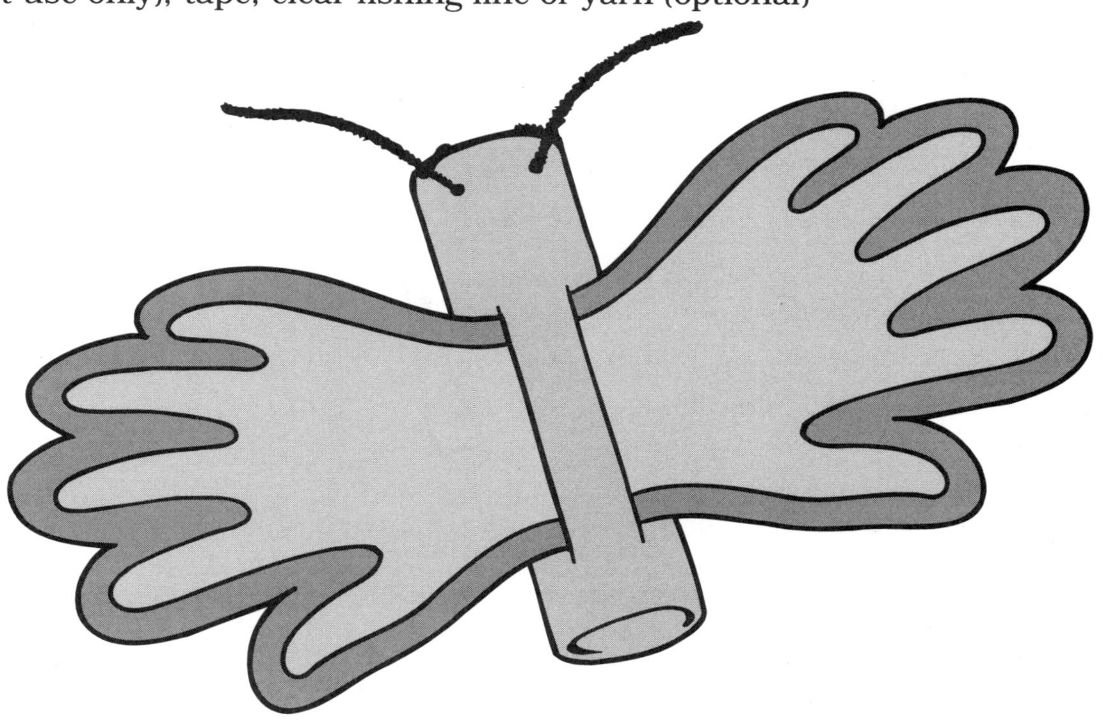

Directions:
1. Cut a slit on opposite sides of each tissue tube. (Each slit should be long enough to insert the base of a child's hand print.) Poke two holes at one end for the antennae.
2. Show the children how to paint a design on the inside of one hand, clap hands together (to make a symmetrical design), and then print their hands onto construction paper.
3. Have the children paint the toilet tissue tubes black.
4. Once the handprints have dried, outline each hand with a marker. Have the children cut out the prints. (These are the wings.)
5. Have the children insert the butterfly wings into the slits in the tube and tape.
6. Loop a pipe cleaner through the two holes in each tube for the antennae.
7. Add butterflies to the Jungle Safari Mural or hang them from the ceiling with clear fishing line or yarn.

Book Link:
- *Where Does the Butterfly Go When It Rains?* by May Garelick (Scholastic)

Stand-Up Tiger

Materials:
Tiger Patterns (p. 35), orange construction paper, black marker or crayon, scissors, glue

Directions:
1. For each tiger's body, fold a 9-inch x 12-inch (23-cm x 30-cm) piece of construction paper in half and draw a line according to the diagram below. Make one body for each child.

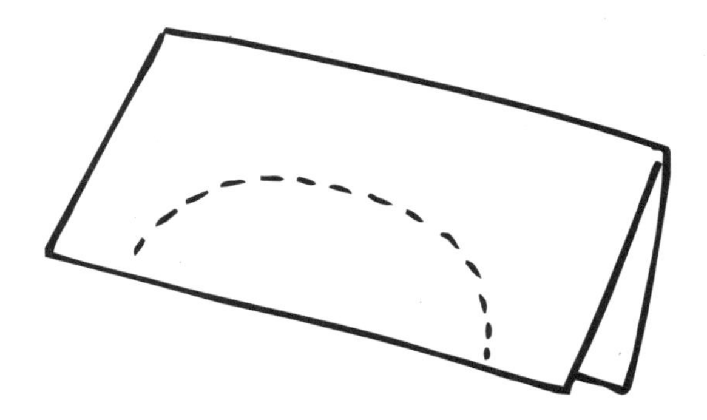

2. Trace the tiger's head, nose, eye, and tail patterns onto construction paper. Make a set for each child.
3. Have the children cut out the patterns.
4. Provide glue for each child to use to attach the head and tail to the body.
5. After they glue on the nose and eyes, give the children markers and crayons to add stripes and other details to their tigers.

Option:
• Glue small strips of black construction paper to the tiger's body for stripes.

Book Link:
• *Sam and the Tigers* by Julius Lester (Dial)

Tiger Patterns

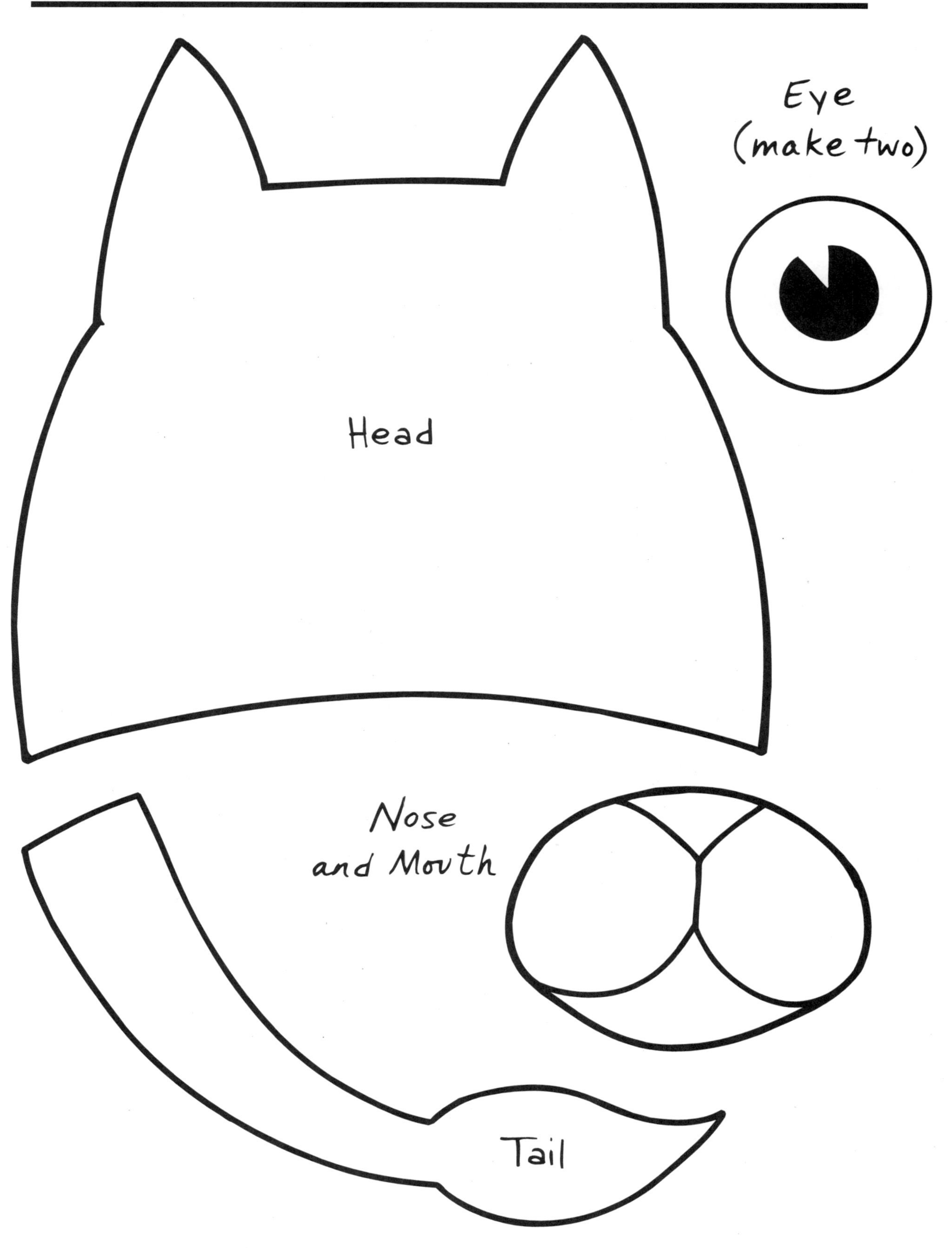

Zany Zebra

Materials:

Zebra Pattern (p. 37), white construction paper, black tempera paint, shallow tins for paint, small toy cars, marker, scissors, newsprint

Directions:

1. Duplicate a copy of the zebra on white construction paper for each child.
2. Cover the workstation with newsprint.
3. Provide toy cars for children to use to roll stripes of paint onto the zebras.
4. Once the paint has dried, cut out the zebras and add them to the Jungle Safari Mural.

Option:

• Instead of using toy cars, children can draw stripes on their zebras using black crayons or markers.

Book Link:

• *Animals Should Definitely Not Wear Clothing* by Judi Barrett (Aladdin)

Zebra Pattern

Coffee-Covered Giraffe

Materials:
Giraffe Pattern (p. 39), yellow construction paper, used dry coffee grounds, shallow tray for coffee grounds, glue, markers, scissors, newsprint

Directions:
1. Enlarge and duplicate a copy of the giraffe and cut it out.
2. Trace the giraffe pattern onto yellow construction paper and cut it out. Make one giraffe per child.
3. Spread the dry coffee grounds in a shallow tray. Place the tray on a newsprint-covered workstation.
4. To make spots on their giraffes, have the children drip glue onto their giraffes, dip the giraffes into the grounds, and blot off the extra grounds.
5. Let the children add details to the giraffe using crayons or markers.
6. Once the giraffes have dried, add them to the Jungle Safari Mural.

Giraffe Pattern

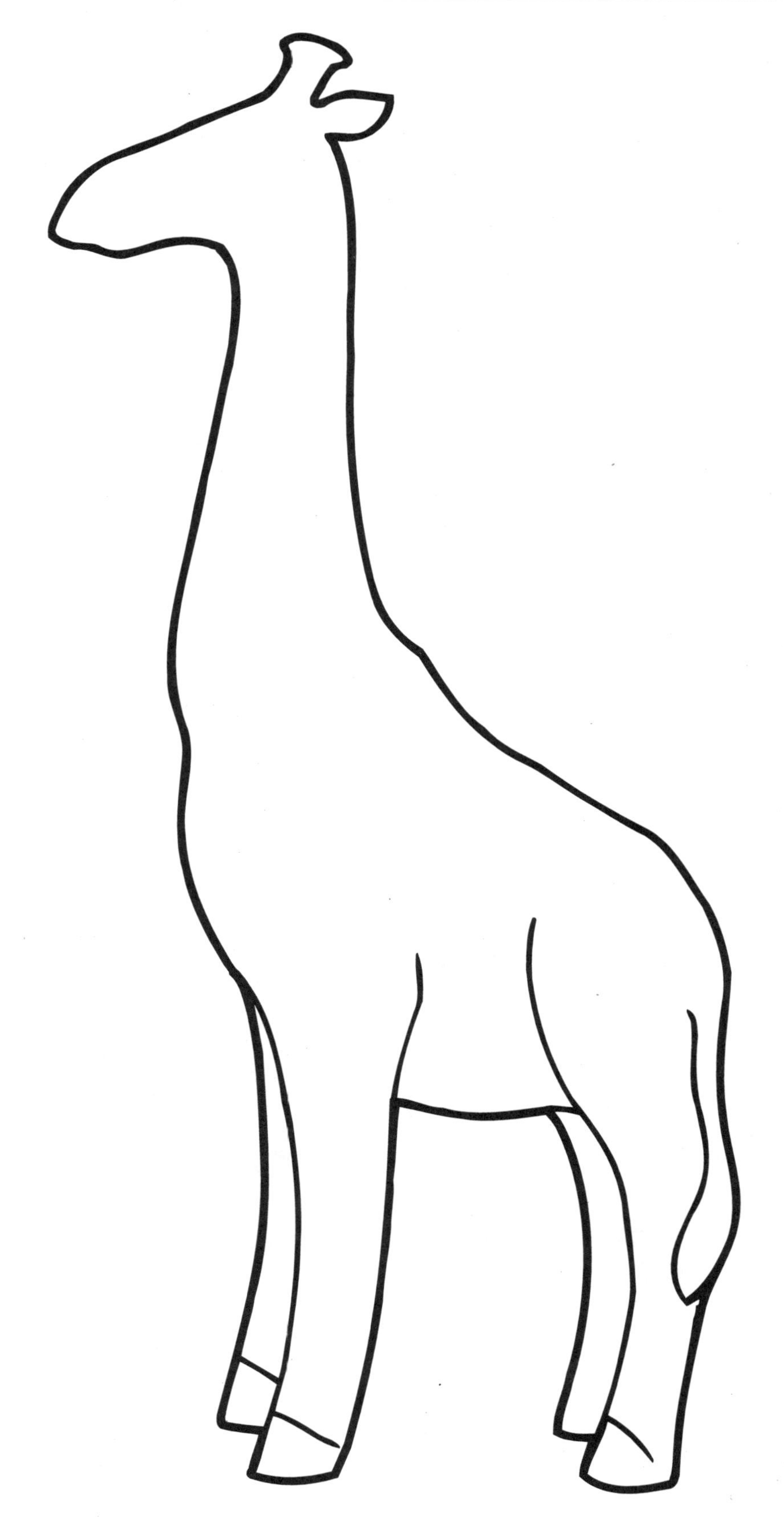

Colorful Fish

Materials:
Fish Pattern (p. 41), sturdy paper, light corn syrup, shallow tins for corn syrup, food coloring, small containers for food coloring, paintbrushes, cotton swabs, scissors, markers, newsprint

Directions:
1. Trace the fish onto sturdy paper and cut it out. Make several for the children to use as templates.
2. Have the children trace fish onto sturdy paper and cut them out.
3. Cover the workstation with newsprint.
4. Have the children brush corn syrup onto their fish.
5. Have the children dip the cotton swabs into different colors of food coloring and dab it onto the corn syrup.
6. Once the corn syrup has dried, add the fish to the Jungle Safari Mural.

Book Link:
• *In the Small, Small Pond*
by Denise Fleming (Henry Holt)

Fish Pattern

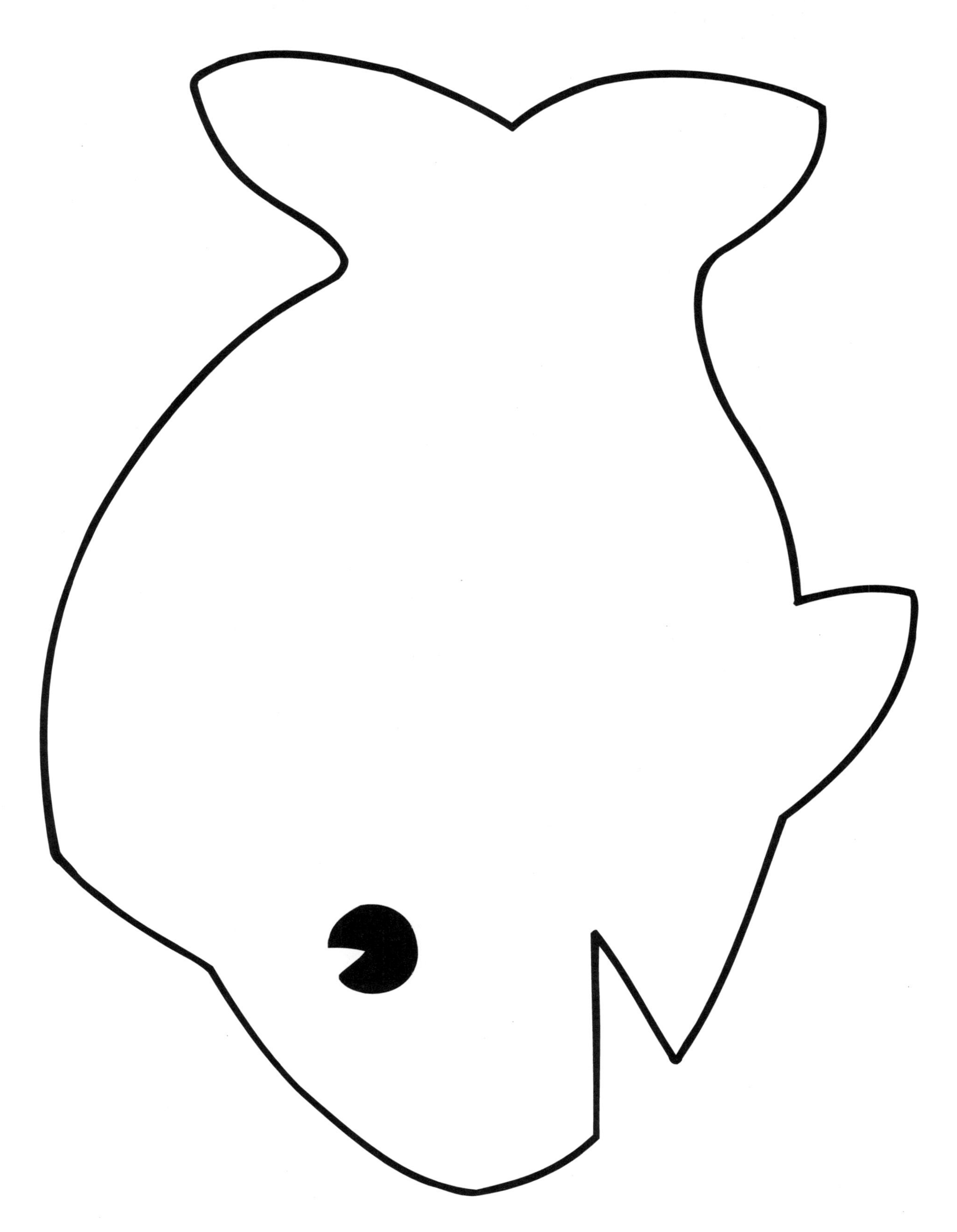

Jungle Grass

Materials:
Empty plastic flat container (ask for one at a nursery or garden shop), plastic trash can liner, potting soil, wheat grass seed, plastic spoons, spray bottles, scissors

Directions:
1. Cut the plastic trash liner to fit the bottom of the flat container. Cut several small holes in the liner.
2. Have the children spoon potting soil into the flat container. They should fill the container half-way.
3. Provide grass seeds for the children to sprinkle onto the soil.
4. Have the children cover the seeds with more potting soil.
5. Give the children spray bottles to water the grass seed daily. Make sure the grass receives sunlight, as well.
6. Have the children observe the progress of their mini-savannah.

Jungle Flannel Board

Materials:
Jungle Patterns (p. 44), flannel board, felt (in assorted colors), scissors

Directions:
1. Trace the Jungle Patterns onto felt and cut them out.
2. Place the three different animals on the flannel board while the children are watching.
3. Have the children close their eyes while you change the sequence of the animals.
4. Have the children open their eyes and try to guess the original sequence of the animals.
5. Repeat this game several times.

Options:
- Let the children play with the flannel board on their own.
- Demonstrate sequencing using the same animal cut from different colors of felt or with the same animal cut in different sizes.

Jungle Patterns

Elephant Number Match

Materials:
Elephant Pattern (p. 46), crayons or markers, self-sticking dots

Directions:
1. Duplicate a copy of the Elephant Pattern for each child.
2. Write the numerals from one to ten on the self-sticking dots. Make a set of dots for each child.
3. Have the children match the numbered dots to the corresponding numerals on the pattern.
4. Provide crayons or markers for the children to use to color their pictures.

Options:
- If self-sticking dots are not available, cut circles from construction paper and number these. Children can glue the circles to the patterns.
- Write the numbers out of order on the dots.
- For more advanced children, white-out the numbers on the pattern before duplicating it. Have the children write the numbers from one to ten in the circles.

Elephant Pattern

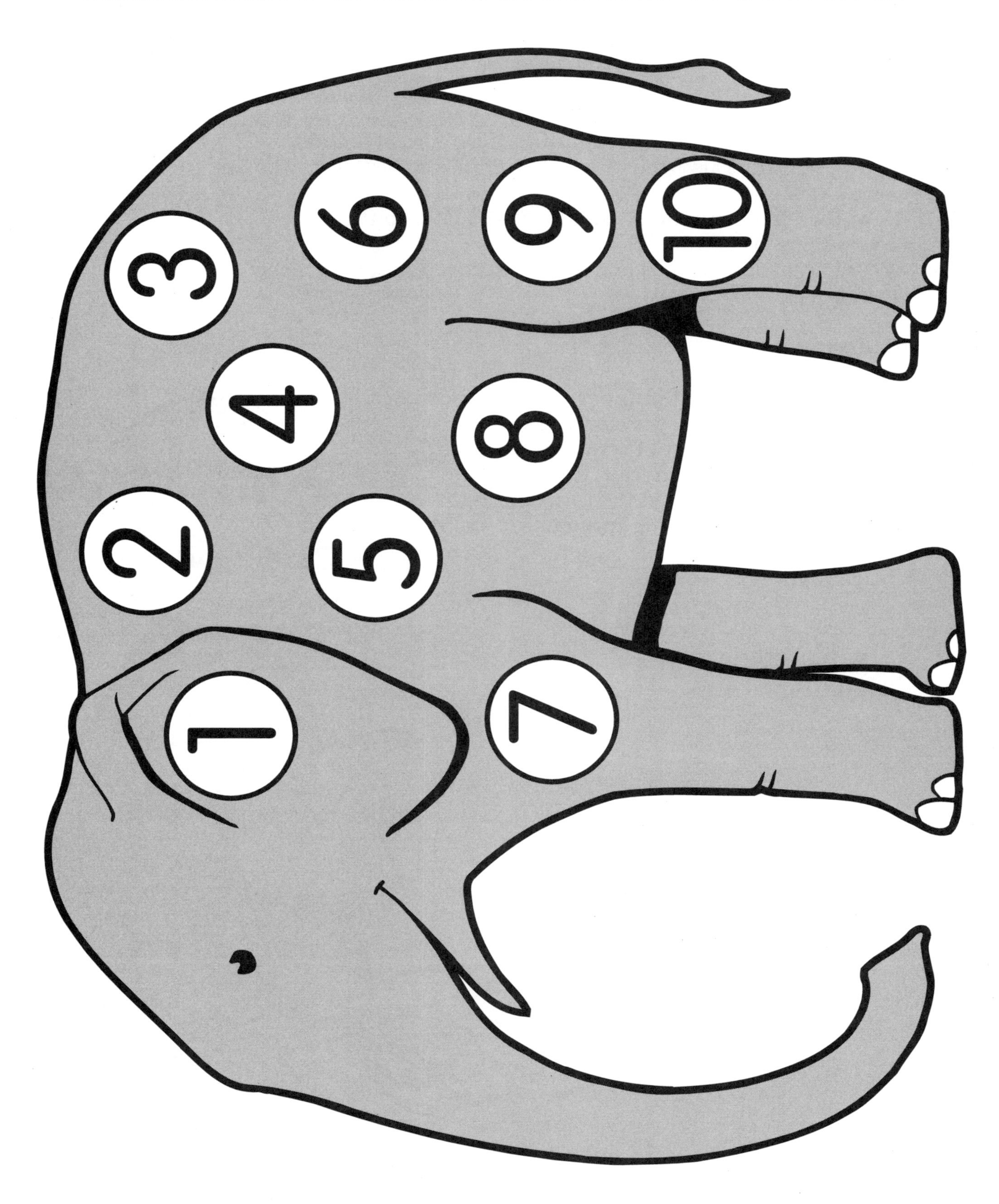

Jungle Memory Match

Materials:

Jungle Patterns (p. 48), colored markers, scissors, clear contact paper

Directions:

1. Duplicate the Jungle Patterns twice, color, cut apart, cover with contact paper or laminate, and cut out again. (Leave a thin laminate border to prevent peeling.)
2. Shuffle the cards and spread them face down on a table.
3. Demonstrate how to play the game. The object is to match the Jungle Patterns by turning the cards over two at a time. If a match is made, the cards remain face up and the child takes another turn. If a match is not made, the cards are turned over and the next child takes a turn. The game continues until all cards are face up.

Options:

- Introduce the game by leaving the shuffled cards face up and having the children simply match the Jungle Patterns together.
- Enlarge and decorate these patterns, and use them to decorate bulletin boards during this unit.

Jungle Patterns